PRAYERS OF THE DESERT

www.ignatius.cc

In association with Legacy Icons
www.legacyicons.com

Printed in the United States

Texts from *The Anthologion* and *The Holy Psalter*,

Edited and designed by Rdr. John Dykstra.
Project oversight by David DeJonge.

ISBN 978-1-7343286-5-3

INTRODUCTION

THE original monks of the Egyptian desert have left us tremendous treasures that speak to us across the centuries. Their survival is a testament to the Spirit of God which inspired them, and their words have much to offer us today.

THE RULE OF SAINT PACHOMIOS (page 5) was revealed to the saint by an angel. After receiving this prayer rule, he maintained it faithfully throughout his life, reciting it at every hour of the day and night. It is a simple prayer rule and ideal for memorization, which can be said in any place or at any time.

THE OFFICE OF THE TWELVE PSALMS (page 11) developed out of the early monastic services, which featured the structured reading of the entire Psalter each day. Over time, other hymns and prayers came to be added to the daily services; but at the same time, the simpler Twelve Psalms service remained as a substitute for Matins, Vespers, and the Lesser Hours. It was prayed especially in small monasteries, or *sketes,* where the proper liturgical resources were not available, or monks did not have the skill or ability to perform the standard services. Note that the Twelve Psalms are not meant to replace the regular prayer rule, but to compliment it, as church services compliment our daily prayers.

In today's world, we see how easily our access to the Church's divine services can be cut off, whether due to pandemic, civil unrest, or violence. God forbid, but we cannot take for granted that we will always be able to attend Church when we desire it. To remain in the life of the Church, we may be required to take on the responsibility of prayer ourselves. In this, we can look to our spiritual ancestors of the desert, who became like angels in their simple lives of constant prayer.

Whether prevented from attending Church due to illness, travel, or other circumstances, or if one simply desires to observe the Church's cycle of prayer in an uncomplicated way, these time-tested prayers have led countless souls on the road to salvation. ☩

THE RULE OF ST. PACHOMIOS

THROUGH the prayers of our holy fathers, O Lord Jesus Christ our God, have mercy on us. Amen.

Glory to you, O God, glory to you.

O Heavenly King, Paraclete, Spirit of Truth, present everywhere, filling all things, Treasury of blessings and Giver of life, come and dwell in us, cleanse us from every stain, and, O Good One, save our souls.

¶ *The prayer* O HEAVENLY KING *is not said from Pascha until Pentecost. But from Pascha until its Leavetaking, say the Paschal Troparion,* CHRIST IS RISEN FROM THE DEAD... *three times in its place.*

THE TRISAGION PRAYERS

Holy God, Holy Strong, Holy Immortal, have mercy on us. **(3)**

Glory to the Father and to the Son and to the Holy Spirit, both now and forever, and to the ages of ages. Amen.

All-holy Trinity, have mercy on us. Lord, forgive our sins. Master, pardon our iniquities. Holy One, visit and heal our infirmities for your Name's sake.

Lord, have mercy. **(3)**

Glory to the Father and to the Son and to the Holy Spirit, both now and forever, and to the ages of ages. Amen.

Our Father in heaven, may your Name be hallowed, your Kingdom come, your will be done on earth as in heaven. Give us today our daily bread. And forgive us our debts, as we forgive our debtors. And do not lead us into temptation, but deliver us from the evil one.

Through the prayers of our holy fathers, O Lord Jesus Christ our God, have mercy on us. Amen.

Lord, have mercy. **(12)**

Glory to the Father and to the Son and to the Holy Spirit, both now and forever, and to the ages of ages. Amen.

Come, let us worship and fall down before the King, our God.

Come, let us worship and fall down before Christ, the King, our God.

Come, let us worship and fall down before him, Christ the King and our God.

PSALM 50

HAVE mercy on me, O God, in your great mercy; according to the fullness of your compassion, blot out my offense.

Wash me thoroughly from my wickedness, and cleanse me from my sin.

For I acknowledge my wickedness, and my sin is ever before me.

Against you only have I sinned and done what is evil in your sight,

That you may be justified in your words, and prevail when you are judged.

For behold, in wickedness I was conceived, and in sin my mother bore me.

For behold, you have loved truth: you have shown me the hidden and secret things of your wisdom.

You will sprinkle me with hyssop, and I shall be cleansed; you will wash me, and I shall be made whiter than snow.

You will make me hear of joy and gladness; the bones which have been humbled will rejoice.

Turn away your face from my sins, and blot out all my iniquities.

Create a clean heart in me, O God, and renew a right Spirit within me.

Do not cast me out from your presence, and do not take your Holy Spirit from me.

Give me back the joy of your salvation, and establish me with a sovereign Spirit.

I will teach transgressors your ways, and sinners will turn to you again.

O God, the God of my salvation, deliver me from bloodshed, and my tongue will rejoice at your righteousness.

O Lord, you will open my lips, and my mouth will declare your praise.

For if you had wanted a sacrifice, I would have given it; you will not take pleasure in burnt offerings.

A sacrifice to God is a broken spirit; a broken and a humbled heart God will not despise.

Do good to Zion, O Lord, in your good pleasure, and let the walls of Jerusalem be rebuilt.

Then you will be well-pleased with a sacrifice of righteousness, oblation and whole burnt offerings.

Then they will offer calves upon your altar.

THE CREED

I BELIEVE in one God, Father almighty, maker of heaven and earth, of all things visible and invisible.

And in one Lord Jesus Christ, the only-begotten Son of God, begotten from the Father before all ages. Light from Light, true God from true God, begotten not made, consubstantial with the Father; through him all things were made; for our sake and for our salvation came down from heaven, and was incarnate from the Holy Spirit and the Virgin Mary and became man; he was crucified also for us under Pontius Pilate, and suffered and was buried; he rose again on the third day, in accordance with the Scriptures, and ascended into heaven and is seated at the right hand of the Father; he is coming again in glory to judge the living and the dead; and his kingdom will have no end.

And in the Holy Spirit, the Lord, the Giver of life; who proceeds from the Father; who together with the Father and the Son is worshiped and together glorified; who spoke through the Prophets.

In One, Holy, Catholic and Apostolic Church. I confess one Baptism for the forgiveness of sins. I await the resurrection of the dead, and the life of the age to come. Amen.

THE JESUS PRAYER

O LORD, Jesus Christ, Son of God, have mercy on me, a sinner. **(100)**

THE DISMISSAL

It is truly right to call you blessed, who gave birth to God, ever-blessed and most pure and the Mother of our God. Greater in honor than the Cherubim, and beyond compare more glorious than the Seraphim, without corruption you gave birth to God the Word, truly the Mother of God, we magnify you.

Glory to the Father and to the Son and to the Holy Spirit, both now and forever, and to the ages of ages. Amen.

Lord, have mercy. **(3)** O Lord, bless.

Through the prayers of our holy fathers, O Lord Jesus Christ our God, have mercy on us. Amen. ☩

THE OFFICE OF THE TWELVE PSALMS

THROUGH the prayers of our holy fathers, O Lord Jesus Christ our God, have mercy on us. Amen.

Glory to you, O God, glory to you.

O Heavenly King, Paraclete, Spirit of Truth, present everywhere, filling all things, Treasury of blessings and Giver of life, come and dwell in us, cleanse us from every stain, and, O Good One, save our souls.

¶ *The prayer* O HEAVENLY KING *is not said from Pascha until Pentecost. But from Pascha until its Leavetaking, say the Paschal Troparion,* CHRIST IS RISEN FROM THE DEAD... *three times in its place.*

THE TRISAGION PRAYERS

Holy God, Holy Strong, Holy Immortal, have mercy on us. **(3)**

Glory to the Father and to the Son and to the Holy Spirit, both now and forever, and to the ages of ages. Amen.

All-holy Trinity, have mercy on us. Lord, forgive our sins. Master, pardon our iniquities. Holy One, visit and heal our infirmities for your Name's sake.

Lord, have mercy. **(3)**

Glory to the Father and to the Son and to the Holy Spirit, both now and forever, and to the ages of ages. Amen.

Our Father in heaven, may your Name be hallowed, your Kingdom come, your will be done on earth as in heaven. Give us today our daily bread. And forgive us our debts, as we forgive our debtors. And do not lead us into temptation, but deliver us from the evil one.

Through the prayers of our holy fathers, O Lord Jesus Christ our God, have mercy on us. Amen.

Lord, have mercy. **(12)**

Glory to the Father and to the Son and to the Holy Spirit, both now and forever, and to the ages of ages. Amen.

¶ *Now continue with the appointed Stases (sections) according to the chart on the next page.*

After saying the usual beginning prayers (pages 11–12), continue with the Stases appointed for the present service:

OUTSIDE LENT		
FOR SERVICE	**STASES**	**PAGES**
Matins	1, 2, 3, 4	14 – 42
The Hours	1, 2	14 – 26
Vespers	3, 4	26 – 42
Compline	3, 4	26 – 42

DURING LENT		
FOR SERVICE	**STASES**	**PAGES**
Matins	1, 2, 3, 4	14 – 42
The Hours	1, 2, 3	14 – 32
Vespers	1, 2, 3	14 – 32
Compline	4	33 – 42

And finally, after completing the appointed Stases, conclude at all times with the Dismissal (page 43).

FIRST STASIS

Come, let us worship and fall down before the King, our God.

Come, let us worship and fall down before Christ, the King, our God.

Come, let us worship and fall down before him, Christ the King and our God.

PSALM 26

THE Lord is my enlightenment and my Savior; whom shall I fear? The Lord is the defender of my life; of whom shall I be afraid?

When evildoers came against me to eat up my flesh, my persecutors and my enemies, they fainted and fell.

Though an army should array itself against me, my heart shall not fear; though war should rise up against me, in this I hope.

I have asked one thing of the Lord, this will I earnestly seek: That I should dwell in the house of the Lord all the days of my life, that I should behold the fair beauty of the Lord, and survey his temple.

For in the day of my afflictions he hid me in his tabernacle, he sheltered me in the secret of his tabernacle, he set me up on a rock.

And now, behold, he has lifted up my head over my enemies.

I went around and offered in his tabernacle the sacrifice of rejoicing; I will sing, I will sing psalms to the Lord.

O Lord, hear my voice which I have raised; have mercy on me, and hear me.

My heart said to you: I have diligently sought your face. Your face, O Lord, will I seek.

Do not turn your face away from me; do not turn away from your servant in anger.

Be my helper; do not forsake me, and do not overlook me, O God my Savior.

My father and my mother have forsaken me, but the Lord has taken me to himself.

Teach me your way, O Lord, and guide me in a right path because of my enemies.

Do not hand me over to the desire of those who torment me, for unjust witnesses have risen up against me, and injustice has lied within herself.

I believe that I will see the good things of the Lord in the land of the living.

Wait on the Lord; be of good courage, and let your heart be strengthened, and wait on the Lord.

PSALM 31

BLESSED are they whose iniquities are forgiven, and whose sins are covered.

Blessed is the one whom the Lord will by no means charge with sin, and in whose mouth there is no guile.

Because I kept silent, my bones grew old from my crying out all day long.

For night and day your hand was heavy upon me, I was turned to wretchedness by a thorn stuck fast in me.

I acknowledged my sin, and did not hide my iniquity. I said: Against myself I will admit my iniquity to the Lord. And you forgave the ungodliness of my heart.

For this, every holy one shall pray to you at a fitting moment. Except in a flood of many waters they will not come near him.

For you are my refuge from the affliction which surrounds me; O my Joy, deliver me from those who have surrounded me.

I will make you understand and guide you in the way in which you should go; I will fix my eyes upon you.

Do not become like horse and mule, who have no understanding; with bit and bridle you must constrain their cheeks so that they do not come near you.

Many are the scourges of the sinner, but mercy will surround the one who hopes in the Lord.

Rejoice in the Lord, and be glad, you righteous, and boast, all you upright in heart.

PSALM 56

HAVE mercy on me, O God, have mercy on me; for my soul has trusted in you.

And in the shadow of your wings will I hope, until iniquity pass by.

I will cry to God Most High, the God who has done good things for me.

He sent from heaven and saved me; he gave to reproach those who trampled on me.

God has sent forth his mercy and his truth, and he has delivered my soul from the midst of lions' cubs; I lay down to sleep, but I am troubled.

As for the sons of men, their teeth are weapons and arrows, and their tongue a sharp sword.

Be exalted, O God, above the heavens, and your glory above all the earth.

They have set a trap for my feet, and have bowed down my soul.

They have dug a pit before me and have fallen into it themselves.

My heart is ready, O God, my heart is ready; I will sing, I will sing psalms.

Rise up, O my glory; rise up, O psaltery and harp; I will rise up early at dawn.

I will confess you, O Lord, among the nations; I will sing to you among the nations.

For your mercy has been magnified even to the heavens, and your truth to the clouds.

Be exalted, O God, above the heavens, and your glory above all the earth.

THE TRISAGION PRAYERS

Holy God. All-holy Trinity. Our Father. Through the prayers.

TROPARIA. TONE 1

Make haste and open to me your fatherly embrace. Like the Prodigal, I have squandered my whole livelihood. As I look to the inexhaustible wealth of your mercy, do not now despise my beggared heart, for to you, O Lord, with compunction I cry: I have sinned against you. Save me!

Glory to the Father...

When you come to earth with glory, O God, the whole world will tremble. The river of fire will flow from your judgment seat, the books will be opened, and the secrets will be revealed. Deliver me then from that unquenchable fire, and count me worthy to stand at your right hand, O Judge most just.

Both now and forever...

All we who with longing flee for refuge to your goodness know you to be Mother of God, and after childbirth still truly Virgin; for we sinners have you as protection; in temptations we have you as salvation, the only All-blameless one.

Lord, have mercy. **(30)**

Glory to the Father and to the Son and to the Holy Spirit, both now and forever, and to the ages of ages. Amen.

SECOND STASIS

Come, let us worship and fall down before the King, our God.

Come, let us worship and fall down before Christ, the King, our God.

Come, let us worship and fall down before him, Christ the King and our God.

PSALM 33

I WILL bless the Lord continually; his praise shall always be in my mouth.

Let my soul boast of the Lord, the humble will hear it and rejoice.

O magnify the Lord with me; let us exalt his Name together.

For I sought the Lord, and he hearkened to me, and he freed me from all my troubles.

Come to him and be enlightened, and your faces shall not be ashamed.

Here is a poor man who cried, and the Lord heard him, and saved him from all his troubles.

The angel of the Lord will encamp round those who fear him, and deliver them.

O taste and see that the Lord is good; blessed is the man who hopes in him!

O fear the Lord, all you his saints, for those who fear him never lack.

The rich became poor and went hungry, but those who seek the Lord shall lack nothing good.

Come, my children, listen to me; I will teach you the fear of the Lord.

Who is there who wants life, who loves to see good days?

Keep your tongue from evil, and your lips from speaking deceit.

Turn away from evil and do good; seek peace and pursue it.

The eyes of the Lord are on the righteous, and his ears are open to their prayer.

The face of the Lord is against those who do evil, to root out their memory from the earth.

The righteous cried, and the Lord heard them, and freed them from all their troubles.

The Lord is close to those who are broken-hearted, and will save the humble in spirit.

The troubles of the righteous are many, but the Lord will deliver them from them all.

The Lord guards all their bones, not one of them will be broken.

The death of sinners is evil, and those who hate the righteous shall go astray.

The Lord will redeem the souls of his servants, and none who hope in him will go astray.

PSALM 38

I SAID: I will guard my ways, that I may not sin with my tongue.

I set a guard on my mouth, while the sinner stood in my presence.

I kept silent and was humbled, and held my tongue, even from good words, and my suffering was renewed.

My heart grew hot within me, and in my meditation a fire would burn.

I spoke with my tongue: O Lord, let me know my end, and the number of my days, what it is, that I may know what I lack.

Behold, you have made my days as a hand's breadth, and my existence is as nothing in your sight.

Indeed, every man alive is altogether vanity.

Surely man passes as a shadow; his turmoil is vanity.

He lays up treasures, but does not know for whom he gathers them.

And now, what is my endurance? Is it not the Lord? My hope is from you.

Deliver me from all my transgressions; you have made me a reproach to fools.

I was silent, and shut my mouth, for you are my Maker.

Remove your scourges from me; I have fainted under the force of your hand.

You rebuke man for sin, and you melt his life away like a spider's web; truly, every man's turmoil is vanity.

Hear my prayer, O Lord, and my supplication; listen to my tears.

Do not be silent, for I am a traveler in the land, and a stranger, as were all my fathers.

Spare me, that I may be refreshed before I depart, and be no more.

PSALM 40

BLESSED is the man who considers the poor and needy; in an evil day the Lord will deliver him.

May the Lord guard him, and preserve his life, and bless him on the earth, and not deliver him into the hands of his enemy.

May the Lord help him on the bed of his pain; in his sickness you have restored him.

I said: O Lord, have mercy on me, heal my soul, for I have sinned against you.

My enemies have spoken evil against me: When will he die, and his name perish?

And though he came to visit me, his heart spoke vanity; he gathered iniquity to himself; he went out, and spoke in the same manner.

All my enemies whispered against me; they plotted evil against me.

They spoke a wicked word against me: Surely one who sleeps will not rise up again.

For even the man at peace with me, in whom I trusted, who ate my bread, has lifted up his heel against me.

But you, O Lord, have mercy on me, and raise me up, and I will repay them.

By this I know that you delight in me, because my enemy shall not rejoice over me.

But you helped me because of my innocence, and have established me in your sight forever.

Blessed is the Lord God of Israel, from everlasting to everlasting. So be it, so be it.

THE TRISAGION PRAYERS

Holy God. All-holy Trinity. Our Father.
Through the prayers.

TROPARIA. TONE 4

Visit my humble soul, O Lord, that has squandered its whole life in sins; receive me as you did the Harlot and save me.

Glory to the Father...

I have wasted my whole life, wretch that I am. O Lord, I cry out like the Prodigal with tearful compunction: O Heavenly Father, I have sinned; cleanse me, and save me! Do not despise me, who have turned away from you, and have become poor because of my fruitless deeds.

Both now and forever...

Now to God's Mother let us humble sinners run in haste and in repentance let us fall down before her feet, crying aloud with fervor from the depths of our souls: O Sovereign Lady, help us now, have compassion upon us; hasten, for we perish from our many offenses. Let not your servants go empty away; we have you as our only hope.

Lord, have mercy. **(30)**

Glory to the Father and to the Son and to the Holy Spirit, both now and forever, and to the ages of ages. Amen.

THIRD STASIS

Come, let us worship and fall down before the King, our God.

Come, let us worship and fall down before Christ, the King, our God.

Come, let us worship and fall down before him, Christ the King and our God.

PSALM 69

O GOD, come to my help; O Lord, attend to help me.

Let those who seek my soul be shamed and confounded.

Let those who desire evil for me be turned back and put to shame.

Let those who say to me, Well done, well done! be turned back immediately, ashamed.

Let all who seek you, O God, be glad and rejoice in you.

Let all who love your salvation ever say: The Lord be magnified.

But I am poor and needy; help me, O God.

You are my helper and my deliverer; O Lord, do not delay.

PSALM 70

IN you, O Lord, have I hoped; let me not be put to shame forever;

In your righteousness deliver me and rescue me; incline your ear to me and save me.

Be a God to defend me, and a stronghold to save me, for you are my foundation and my refuge.

Deliver me, O my God, from the hand of the sinner, from the hand of the transgressor and wrongdoer.

For you are my endurance, O Lord; O Lord, you are my hope from my youth.

On you have I leaned from the womb; from my mother's womb you are my protector.

Of you is my unending song of praise; I have become as a wonder to many, and you are my strong helper.

Let my mouth be filled with your praise, that I may hymn your glory and your majesty all the day long.

Do not cast me off in my old age; when my strength fails, do not forsake me.

For my enemies have spoken against me, and those who lay in wait for my soul have plotted together, saying: God has forsaken him; persecute him and take him, for there is no one to deliver him.

My God, do not be far from me; my God, attend to help me.

Let those who plot against my soul be ashamed and utterly fail; let those who seek my hurt be clothed with shame and dishonor.

But as for me, I will hope continually, and will praise you more and more.

My mouth will declare your righteousness openly, and your salvation all the day long; for I am not acquainted with the affairs of men.

I will go on in the might of the Lord; O Lord, I will make mention of your righteousness alone.

O my God, you have taught me from my youth, and I will still proclaim your wonders.

Even to old age and advanced years, O my God, do not forsake me,

Until I proclaim your mighty arm to every generation that is to come,

Even your power and your righteousness, O God, up to the highest heavens, even the mighty works that you have done. O God, who is like you?

How great are the afflictions, many and evil, you have shown me! Yet you turned and enlivened me, and brought me up again from the depths of the earth.

You multiplied your righteousness, and turned and comforted me, and brought me again out of the depths of the earth.

Therefore will I confess you, O God, because of your truth, on an instrument of psalmody; I will sing psalms to you on the harp, O Holy One of Israel.

My lips will rejoice when I sing to you, and my soul which you have redeemed.

And my tongue will also meditate on your righteousness all the day long, when they will be ashamed and confounded that seek evils for me.

PSALM 76

I CALLED to the Lord with my voice, with my voice to God, and he heard me.

In the day of my trouble, I sought the Lord with my hands lifted up by night before him, and I was not deceived.

My soul refused to be comforted; I remembered God, and I was glad; I spoke idly, and my soul grew faint.

My eyes were wakeful and kept watch; I was troubled and did not speak.

I considered the days of old, and remembered ancient years, and I meditated.

I pondered in my heart by night, and diligently searched my spirit, saying:

Will the Lord cast me off forever, and will he find favor no more?

Will he cut off his mercy forever, from generation to generation?

Will God forget to have mercy? Or will he shut up his compassions in his wrath?

And I said: Now have I begun; this change is of the right hand of the Most High.

I have remembered the works of the Lord; for I will remember your wonders from the beginning.

And I will meditate on all your works, and I will consider your ways.

O God, your way is in the sanctuary. What God is great as our God? You are God, who alone works wonders.

You have made your power known among the peoples; with your arm you have redeemed your people, the sons of Jacob and Joseph.

The waters saw you, O Lord, the waters saw you and were afraid; and the depths were troubled.

There was a great roaring of waters, the clouds raised a voice.

For your arrows went abroad; the voice of your thunder rolled.

Your lightnings flashed in the world; the earth quaked and trembled.

Your way is in the sea, and your paths in many waters; and your footsteps cannot be known.

You guided your people as sheep by the hand of Moses and Aaron.

THE TRISAGION PRAYERS

Holy God. All-holy Trinity. Our Father. Through the prayers.

TROPARIA. TONE 6

I think of the fearful Day, and weep over my filthy deeds. What answer shall I give to the immortal King? And how shall I, the prodigal, have boldness to look upon the Judge? Have mercy upon me, O compassionate Father, only-begotten Son, and Holy Spirit.

Glory to the Father...

When you sit in righteous judgment, O Merciful, in the valley of tears, in the place you devised, do not make public my secret acts. Do not put me to shame before the Angels, but spare me and have mercy on me, O God.

Both now and forever...

Open the gate of compassion to us, blessed Mother of God; hoping in you, may we not fail. Through you, may we be delivered from adversities, for you are the salvation of the Christian race.

Lord, have mercy. **(30)**

Glory to the Father and to the Son and to the Holy Spirit, both now and forever, and to the ages of ages. Amen.

FOURTH STASIS

Come, let us worship and fall down before the King, our God.

Come, let us worship and fall down before Christ, the King, our God.

Come, let us worship and fall down before him, Christ the King and our God.

PSALM 101

O LORD, hear my prayer, and let my cry come to you.

Do not turn your face from me in the day I am in tribulation. Incline your ear to me; in the day I call upon you, be swift to hear me.

For my days have vanished like smoke, and my bones have been burnt up like brushwood.

I have been smitten like grass, and my heart has been dried up, so that I have forgotten to eat my bread.

My bones have stuck to my flesh, from the sound of my groaning.

I have become like a pelican of the desert; like an owl among ruins.

I have kept vigil, and become like a sparrow, alone upon a house-top.

All day my enemies have reviled me, and those who praised me have sworn an oath against me.

I have eaten ashes as my bread, and mixed my drink with weeping.

From the face of your wrath and your anger; for having lifted me up, you have cast me down.

My days have declined like a shadow, and I have been dried up like grass.

But you, O Lord, abide forever, and your memorial to generation and generation.

You will arise and have mercy on Zion, for it is time to have mercy on her, for the time has come.

For your servants have been well-pleased with her stones, and they will have mercy on her dust.

And the nations will fear your Name, O Lord, and all the kings of the earth your glory.

For the Lord will build up Zion, and appear in his glory.

He has looked on the prayer of the humble, and has not despised their supplication.

Let this be written for another generation, and a people that is being created will praise the Lord.

For the Lord has leaned down from his holy height, he has looked from heaven upon the earth,

To hear the groaning of those in fetters, to free the children of the slain,

To announce the Name of the Lord in Zion and his praise in Jerusalem.

When peoples are gathered together, and kings to serve the Lord.

He answered him in the way of his strength. Tell me the fewness of my days.

Do not lead me away halfway through my days. Your years are for generations of generations.

In the beginning it was you, O Lord, who laid the foundations of the earth, and the heavens are the works of your hands.

They will perish, but you endure; they will all grow old as a garment;

As a mantle you will roll them up, and they will be changed; but you are the same, and your years will not fail.

Your servants' children shall have their dwelling, and their seed shall be guided forever.

THE PRAYER OF MANASSEH

O LORD Almighty, the God of our Fathers, of Abraham and Isaac and Jacob and of their righteous seed;

Who made the heaven and the earth with all their array; who shackled the sea by the word of your command;

Who shut up the deep and sealed it with your dread and glorious Name;

Before whom all things shudder and tremble in the presence of your power;

For the majesty of your glory cannot be borne, and the wrath of threat to sinners is irresistible,

And the mercy of your promise is measureless and unsearchable.

For you are the Lord Most High, compassionate, long-suffering and full of mercy, and you repent over the evils of mankind.

You, O Lord, according to the multitude of your goodness, have appointed repentance and forgiveness to those who have sinned against you,

And in the multitude of your mercies you have decreed repentance unto salvation for sinners.

You therefore, O Lord God of Powers, did

not appoint repentance for the righteous, for Abraham and Isaac and Jacob, who did not sin against you,

But you have appointed repentance for me, a sinner, for I have sinned above the number of the sand of the sea.

My iniquities have been multiplied, O Lord, my iniquities have been multiplied, and I am not worthy to raise my eyes and to see the height of heaven for the multitude of my unrighteous deeds.

I am bowed down by a heavy iron fetter, so that I cannot lift my head, and there is no respite for me;

For I have provoked your wrath and done what is evil in your sight, not doing your will nor keeping your commands.

And now I bow the knee of my heart, praying for the goodness which is from you:

I have sinned, O Lord, I have sinned, and I recognize my iniquities; but I ask with supplication:

Forgive me, O Lord, forgive me, and do not destroy me with my iniquities,

Do not be wrathful forever; do not lay up evils for me, nor condemn me to the lowest parts of the earth:

For you are God, the God of the penitent, and in me you will show all your goodness;

For you will save me who am unworthy, according to your great mercy, and I will praise you continually all the days of my life.

For all the powers of heaven praise you, and yours is the glory to the ages of ages. Amen.

THE LITTLE DOXOLOGY

GLORY to God in the highest, and on earth peace, goodwill among men.

We praise you, we bless you, we worship you, we glorify you, we give you thanks for your great glory.

O Lord, King, God of heaven, Father Almighty: O Lord, Only-begotten Son, Jesus Christ and Holy Spirit.

O Lord God, Lamb of God, Son of the Father, who take away the sin of the world, have mercy on us; you take away the sins of the world.

Receive our prayer, you who sit on the right hand of the Father, and have mercy on us.

For you alone are holy, you alone are Lord, O Jesus Christ, to the glory of God the Father. Amen.

Every day will I bless you, and I will praise your Name forever, and to ages of ages.

O Lord, you have been our refuge from generation to generation.

I said: O Lord, have mercy on me, heal my soul, for I have sinned against you.

O Lord, I have run to you for refuge; teach me to do your will, for you are my God.

For with you is the source of life, and in your light shall we see light.

O continue your mercy toward those who know you.

Grant, O Lord, this day to keep us without sin.

Blessed are you, O Lord, the God of our fathers, and praised and glorified is your Name to the ages. Amen.

Let your mercy, O Lord, be upon us, as we have hoped in you.

Blessed are you, O Lord, teach me your statutes.

Blessed are you, O Master, make me understand your statutes.

Blessed are you, O Holy One, enlighten me with your statutes.

Your mercy, O Lord, is forever; do not reject the work of your hands.

To you praise is due, to you song is due, to you glory is due: to the Father and to the Son and to the Holy Spirit, both now and forever, and to the ages of ages. Amen.

THE PRAYER OF SAINT EUSTRATIOS

I MAGNIFY you greatly, O Lord, because you have looked upon my lowliness, and have not hemmed me into the hands of enemies, but have saved my soul from constraints. And now, O Master, let your hand protect me, and your mercy come upon me, for my soul has been troubled and is greatly afflicted at its departure from this wretched and soiled body of mine. May the evil plan of the adversary never confront and obstruct it, because of the many sins committed by me in this life in knowledge and in ignorance. Be merciful to me, O Master, and never let my soul behold the dark and gloomy sight of the evil demons; but may your bright and shining Angels receive it. Give glory to your holy Name, and bring me by your power to your divine judgment seat. When I am judged, let not the hand of the ruler of this world seize me to cast me, sinner that I am, into the depths of Hades; but stand by me and be for me a savior and a helper. Have mercy, O Lord, on my soul, stained with the passions of life, and receive it pure through repentance and confession; for you are blessed to the ages of ages. Amen.

THE TRISAGION PRAYERS

Holy God. All-holy Trinity. Our Father. Through the prayers.

TROPARIA. TONE 8

With your compassionate eye, O Lord, behold my humiliation, for little by little my life is being squandered, and there is no salvation for me from works; therefore I ask this: With your compassionate eye, O Lord, see my humiliation and save me.

Glory to the Father...

My time has come to an end, and your dread tribunal is ready. My life passes by, and judgment awaits me; my sentence is fiery torment and the unquenchable flame. But grant me a fountain of tears; quench the inferno by your might, for you desire the salvation of all.

Both now and forever...

You were born for us from a Virgin, and endured crucifixion, O loving Lord; by your death you despoiled Death, and revealed resurrection as God; do not despise those whom you fashioned with your own hand; show us your compassion, O Merciful; accept the Mother of God who bore you as she intercedes for us, and save, O our Savior, a people in despair.

Lord, have mercy. **(30)**

Glory to the Father and to the Son and to the Holy Spirit, both now and forever, and to the ages of ages. Amen.

¶ *Now conclude with the Dismissal (next page).*

THE DISMISSAL

Greater in honor than the Cherubim, and beyond compare more glorious than the Seraphim, without corruption you gave birth to God the Word, truly the Mother of God, we magnify you.

Glory to the Father and to the Son and to the Holy Spirit, both now and forever, and to the ages of ages. Amen.

Lord, have mercy. **(3)** O Lord, bless.

O Lord Jesus Christ, Son of God, through the prayers of your all-pure and holy Mother; by the power of the precious and life-giving Cross; through the protection of the honored, Bodiless Powers of heaven; by the intercessions of our venerable and God-bearing fathers; of the holy Prophet David; and of all the Saints, have mercy on me and save me, a sinner, for you are good and love mankind.

Through the prayers of our holy fathers, O Lord Jesus Christ, our God, have mercy on us. Amen. ✝

GLORY TO GOD FOR ALL THINGS. AMEN